Calling all social entrepreneurs!

Hello, I'm Jonah.

I taught myself to crochet when I was five years old, and since then I have started a business called Jonah's Hands which has a goal of bringing the world together, one stitch at a time.

My parents adopted me from Ethiopia when I was six months old and I was given opportunities that my brothers and sisters in Ethiopia were not. So I decided to make a better world. I teamed up with Roots Ethiopia to build a library and a science centre in the region where I was born using the proceeds of my crocheted items.

The children in this book teach us that when each of us takes action, we can help to resolve social problems or work to save the planet.

You don't have to wait until you are a grown-up to start making a social impact. You could do something close to home or — as in my case — halfway around the globe. The future of our world begins with you... today!

JONAH LARSON
Social entrepreneur and founder of Jonah's Hands

In this book, meet 12 real-life children...

Maya from the USA
Fighting fast fashion

Lena from Germany
Working to end period poverty

Reyhan from Azerbaijan
Promoting green energy

IT'S OUR BUSINESS to MAKE A BETTER WORLD

Written by **Rebecca Hui**
Illustrated by **Anneli Bray**
Foreword by **Jonah Larson**

MAGIC CAT PUBLISHING

Fabiënne from the Netherlands

Reducing paper waste

Aahan and Amal from the UK

Putting a stop to single-use straws

Archie from Australia

Ambassador of the Seabin Project

Chmba from Malawi

Supporting women and girls through arts and education

Jiahua from China

Redistributing food destined for landfill

Jefferson from Kenya

Growing food through sustainable farming

Thomas from France

Protecting our oceans

Chaeli from South Africa

Campaigning for disability access

Omid from Italy

Making healthcare more accessible

End the trend for fast fashion...

Maya from the USA was concerned about the environmental impact of 'fast fashion' – when cheap clothes are made very quickly, only to be worn for a short time and thrown away. So, at eight years old, she launched eco-fashion brand Maya's Ideas to sell clothes made from organic, recycled and vintage materials. She also spends her time educating others about the impact that the fashion industry has on the planet.

Cheap clothes can often mean that the way they were made was harmful to the environment, such as the use of synthetic materials, like polyester and acrylic.

Globally, three out of five items of clothing end up on a landfill site within a year of buying them.

Reduce paper waste...

Fabiënne from the Netherlands was roused to action after watching videos of forests being cut down all over the world. When she was seventeen years old, she started Grow a Wish and began to sell greetings cards made from recycled paper that contain a surprise inside: seeds! Instead of throwing the greetings card away, it can be replanted and grown into basil, tomatoes or summer flowers to help the environment.

Fabiënne sells a greetings card that will grow into a beautiful plant.

Around the world, an estimated 4 billion trees are cut down every year for the making of paper products.

NAME: Fabiënne Overbeek
COUNTRY: Netherlands
CHANGEMAKER FOR: Creating greetings cards which can be planted rather than discarded

Greetings cards can feature embellishments such as ribbons, glitter or foil which cannot be recycled.

We live in a society which uses products for a short time and then disposes of them, known as a throwaway culture.

Plants clean our water, soil and air — making our environment a healthier place to live.

By growing your own garden, you can decide what goes into the soil, preventing harmful chemicals from polluting our environment.

Say no to single-use plastic...

After seeing plastic straws littering the ocean while on holiday, British brothers Aahan, thirteen, and Amal, seventeen, decided to set up their own straw business, **The Last Straw Cheltenham.** They sell bamboo and wheat straws to shops, restaurants and cafes as a sustainable alternative to single-use plastic straws. They also campaign to inform others of the devastating effects of plastic on the oceans.

Bamboo is the world's fastest-growing plant. It is an environmentally friendly alternative to plastic as it is durable and compostable.

In most places, plastic straws cannot be recycled and many end up in the oceans.

The best way to fight plastic is to avoid buying it in the first place.

End period poverty...

Lena, from Germany, first travelled to Namibia when she was three years old. A little over a decade later, she returned and saw how important it is for children to have access to education, but without access to period products, many girls have to stay at home. So she set up an organization, Wadadee Cares, dedicated to keeping children in school, and one of its projects, NamPads, provides sustainable, reusable sanitary towels made by local seamstresses.

Education is a basic human right for all.

Periods are a natural process and a part of many girls' lives.

Period poverty means being unable to access sanitary products due to the cost.

$1 + 1 = 2$
$2 + 2 = 4$
$3 + 3 =$

NAME: Lena Palm
COUNTRY: Germany
CHANGEMAKER FOR: Providing sanitary products to keep girls in school

The stigma surrounding periods has been shown to directly affect a girl's potential to succeed.

Lena hands fabric to a local seamstress.

In the United Kingdom, an estimated 49 per cent of girls have missed a day of school due to their periods.

Share skills to help others...

When artist Chmba found out that girls whose families lived in poverty were forced to leave school early, she developed a sustainable solution, even though she was only sixteen at the time. She launched Tiwale and trained women and girls to dye-print fabrics to sell in their home country of Malawi. The money made from sales provided grants for those interested in going back to school. So far, Chmba has helped over 300 women and girls, with each one keeping 60 per cent of their profits and giving 40 per cent back for the future training of others.

When young girls face barriers to education early in life, it is more difficult for them to get a job when they are older.

Around the world, nearly one in four girls between the ages of fifteen and nineteen are neither employed nor in education or training.

Globally, education increases earnings by up to 10 per cent for each additional year in school.

Chmba teaches others to dye-print fabrics.

Women living in the world's poorest countries are less likely to get the education they need or bank the money they earn.

NAME: Chmba Ellen Chilemba
COUNTRY: Malawi
CHANGEMAKER FOR: Supporting women and girls through art and education

Equal rights and opportunities helps everyone to fulfil their potential.

TIWALE

Provide food to those in need...

When Jiahua learned that too much food was being thrown away in her home country of China, she decided to make use of the food waste. At the age of seventeen, she set up PDT Food Depot to rescue food that would end up in landfill and give it to people in need. She has collected more than 30 tonnes of food from supermarkets, farms and factories, and redistributed it to forty-eight communities so far.

Food that is fit for humans to eat but isn't eaten is called food waste. Food may be left to go off or is thrown away because it has past its use-by date or is no longer wanted.

Over one-third of all food produced is lost or wasted every year globally.

Jiahua helps a community member to select their food for the week.

Grow your own food...

Jefferson was six years old when he realized that food security was an issue for his community in his home country of Kenya. The local farms were hit by a drought which affected the crops, leaving his community hungry. Ten years later, he started a sustainable food system and grew hydroponic tomatoes indoors. His eco-friendly business, Eden Horticultural Hub, runs four hydroponics systems to supply tomatoes to more than a hundred households and lunches to 1,500 school pupils.

A sustainable food system delivers food security and nutrition for all.

Use green energy...

Reyhan was just fourteen years old when she designed a smart device for green energy: to generate electricity from raindrops. She came up with the idea in her home country of Azerbaijan, after thinking that rather than rainwater disrupting their electrical supply, it could perhaps create it instead! Her rainwater collector fills the water tank with rainwater that flows at high speed through the electric generator to produce energy. When there is no rain, the energy can still be stored in batteries. To promote her invention, Reyhan founded a company named Rainergy with the motto, 'Light up one house at a time.'

Green energy is any energy type that is generated from natural resources, such as sunlight, wind or water.

Reyhan monitors the collection of rainwater.

The Rainergy device eases the pressure on the local power grid by giving communities an extra source of electricity.

Green energy sources are naturally replenished, as opposed to fossil fuel sources, like natural gas or coal, which can take millions of years to develop.

When it rains, billions of litres of water can fall and the volume has enormous electric potential.

Green energy sources release less carbon dioxide from the environment than natural gas or coal-based energy.

NAME: Reyhan Jamalova
COUNTRY: Azerbaijan
CHANGEMAKER FOR: Green energy

Educate others to protect our oceans...

Fourteen-year-old Thomas from France knew how important it was to protect our oceans and so he set up a boat school to educate others. He sailed around the Atlantic Ocean, visiting schools to inform children about the harmful effects of acidification and overfishing. His organization, Children for the Oceans, is now a worldwide community of ambassadors who provide educational materials on ocean conservation.

The oceans are becoming more acidic because of the extra carbon dioxide in the atmosphere caused by humans cutting down forests and burning fossil fuels.

Overfishing is when people catch fish faster than it can be replenished. This can affect the ecosystem and disrupt the food chain.

Champion a more inclusive world...

Diagnosed with cerebral palsy at eleven months old, Chaeli from South Africa has been a wheelchair user her whole life. At the age of nine, Chaeli, along with her sister and three friends, started a fundraiser to buy a motorized wheelchair to provide Chaeli with more freedom and independence. This fundraiser was the start of The Chaeli Campaign, a social justice organization co-founded to promote and provide the mobility and educational needs of children with disabilities.

Between 93 million and 150 million children live with a disability worldwide.

Chaeli explains how a motorized wheelchair gives her independence.

In some countries, people have access to healthcare whenever they need it, but for others this is not an option.

Some people don't get the healthcare services they need because they don't have enough money or they live too far away from providers who offer them.

Having access to healthcare is considered to be a basic human right.

Build a community of caregivers...

When fifteen-year-old Omid's best friend lost his father due to heart failure, he decided to act. He founded Aid You (Mobile Cardiologist) in his home country of Italy, which provides affordable and accessible medical devices to monitor patients in need. The community-based initiative trains local volunteers to give them first-aid skills to help patients in times of emergencies. Omid also teamed up with a co-founder from Uganda to provide the same services throughout East Africa.

How can you help to build a more sustainable world?

Addressing the social and environmental challenges facing our planet will require a generation of social innovators and changemakers. Social entrepreneurs are agents of positive change who address challenges through an enterprising approach. They develop businesses that trade for a social or environmental purpose, reinvest profits into their mission and are accountable for their actions. They combine insight, compassion and imagination to create a better, fairer world.

1. **Find your passion.** Notice what is making you sad, angry or excited, and transform that energy into creativity.

2. **Tackle one problem at a time.** Think about what you can do now — today. Think small, then dream big.

3. **Have a clear social mission.** Start simple first and let your values guide your journey.

4. **Think global, act local.** Even if your cause is a local one, look beyond to see what others have done around the world.

5. **Keep prototyping.** Keep trying different things and do not be afraid to fail. Failure simply means there is something to be learned.

6. **Do research.** Learn more about the field you are interested in and who the key players are.

7. **Involve others.** A project grows when others can get involved, so engage with your community for help.

8. **Stay up to date.** Watch, listen and read as much as you can to stay informed about the world around you.

9. **Keep at it.** Being entrepreneurial is hard work. Being able to stay motivated and having belief in yourself and your ideas is key.

10. **Tell your story.** Be confident and talk to your friends and family about why you're making a social impact.

Ten things you can do to be a responsible consumer:

1. Buy less. Before you decide to buy or not, ask yourself if you really need it. We can live without many things and we need less than we usually buy.

2. Buy quality over quantity. Instead of buying five T-shirts that you'll wear once and need to replace, you could buy one high-quality T-shirt to last a lifetime.

3. Be selective. Opt for brands and shops that pride themselves on their quality and sustainable values.

4. Upcycle an old item. Instead of throwing something away, repair, decorate or change it so that it can be used again as something more valuable.

5. Read and question labels. If a product is sustainable, it will always indicate this on the label. Also do your own research to make sure the product was not harmful to the environment.

6. Use green energy. Switch electrical items off, turn the heating down and consider asking your family to change your energy provider to a greener alternative.

7. Avoid plastic packaging. Opt for reusable items, such as carrying your own reusable bag or water bottle.

8. Stop food waste. Keep track of the food you've bought to avoid throwing any away.

9. Consume seasonal produce. Buy seasonal fruits and vegetables, grown in local places, to avoid the promotion of intensive horticultural productions in other areas, which demand large amounts of water.

10. Be aware. No one can be a perfect consumer, but by becoming more aware of ethical practices, we can start to hold companies to a higher standard.

Further reading

With the help of an adult, find out more about the children and issues featured in this book using these websites:

entrepreneurship-campus.org
globalcitizen.org
peacefirst.org
wearefamilyfoundation.org
mayasideas.com
growawish.nl
thelaststrawcheltenham.co.uk
seabinproject.com
wadadeecares.com
tiwale.org
childrenfortheoceans.eu
chaelicampaign.org

REBECCA HUI is an artist and social entrepreneur. She studied Business Administration at the University of California, Berkeley, USA, and City Planning, City Design and Development at MIT before starting Roots Studio, a company which reimagines cultural preservation by building bridges between rural communities and the global fashion market. Rebecca is driven by an ethos of sustainability, and is a Forbes 30 Under 30, Cartier Women's Initiative Laureate and Echoing Green Fellow.

ANNELI BRAY is a children's illustrator from North West England. From an early age she could be found painting stories about animals and other magical creatures, making books about ponies and reading voraciously. Anneli graduated with a First in Illustration from Norwich University of the Arts, became a bookseller (drawing on breaks) and at night dreamed of having her own illustrations in a book one day. Now she is best known for her warm, colourful illustrations that are influenced by her love of travel, nature and adventure.

JONAH LARSON is a crochet prodigy from Wisconsin, USA. Jonah began crocheting at five years old and has been making a social impact ever since. He donates his crocheting proceeds and has built a library and science centre in the village in Ethiopia from where he was adopted. Jonah also runs his own business, Jonah Hands, where he features crochet tutorials. In 2021, Jonah was named Young Entrepreneur of the Year in his home state, and his infectious desire to give back has earned him a worldwide following.

MAGIC CAT PUBLISHING

It's Our Business to Make a Better World © 2022 Lucky Cat Publishing Ltd
Text © 2022 Rebecca Hui • Foreword by Jonah Larson
Illustrations © 2022 Anneli Bray
First Published in 2022 by Magic Cat Publishing, an imprint of Lucky Cat Publishing Ltd,
Unit 2 Empress Works, 24 Grove Passage, London E2 9FQ, UK
This paperback edition first published 2023

The right of Rebecca Hui to be identified as the author of this work and Anneli Bray to be identified as the illustrator of this work has been asserted by them in accordance with the Copyright, Designs and Patents Act, 1988 (UK).

All rights reserved.

No part of this publication may be reproduced, stored in a retrieval system, or transmitted, in any form, or by any means, electrical, mechanical, photocopying, recording or otherwise without the prior written permission of the publisher or a licence permitting restricted copying.

A catalogue record for this book is available from the British Library.

ISBN 978-1-913520-88-5

The illustrations were painted using gouache, coloured pencils and digital media • Set in Rainer and Panforte Pro

Published by Rachel Williams and Jenny Broom
Designed by Nicola Price • Edited by Helen Brown

The publication is not authorised, licensed or approved by any of the children featured in this book.

Manufactured in China.

9 8 7 6 5 4 3 2 1